A Sea of New Media

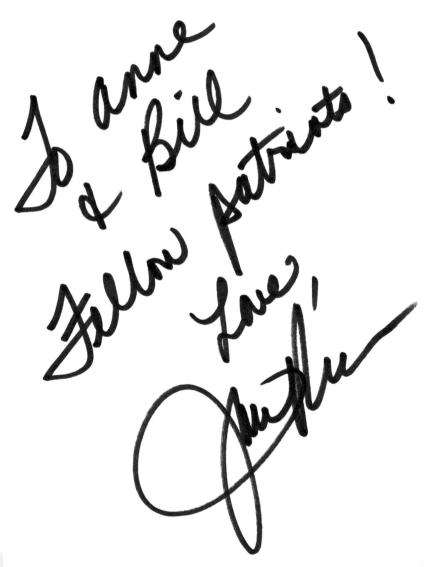

To Anne & Bill
Fellow Patriots!
Love,

A Sea of New Media

TRANSFORMATION OF
THE AMERICAN PRESS

• • •

Dr. Jane Ruby

Copyright © 2017 Dr. Jane Ruby

All rights reserved. This book or any portion thereof may not be reproduced or used in any manner whatsoever without the express written permission of the publisher except for the use of brief quotations in a book review.

Publishing services provided by:

ISBN-13: 9781548258832
ISBN-10: 1548258830

Dedication

This is dedicated to all of the New Media warriors famous and not yet famous who have inspired me and in the path laid by Andrew Breitbart, are restoring the First Amendment of the United States Constitution, Freedom of the Press - and the preservation of liberty by a populace that is *truthfully* informed.

And to my mother, Alfie Piampiano who gave me my integrity and my guts.

Contents

	Acknowledgements · · · · · · · · · · · · · · · · · xiii	
Introduction	How I Was Stung By The Fake News Media at 16· ·xv	
Chapter 1	A Sea of New Media · · · · · · · · · · · · · · · ·	1
Chapter 2	Many Perfect Storms· · · · · · · · · · · · · · · ·	9
Chapter 3	Old Corrupt Media: How Did We Get Here? ·	25
Chapter 4	New Media: What Does It Look Like?	·35
Chapter 5	The Equalizers: Who are the New Media Warriors? · · · · · · · · · · · · · · · · · ·	53
Chapter 6	Suicide Watch: Implosion of the Mainstream Media · · · · · · · · · · · · · · · · ·	65
Chapter 7	New Media Rising: It's About The Movement ·	73

Acknowledgements

To every person along my path who told me anything was possible and to Mike Cernovich and Jack Posobiec for pioneering citizen journalism, role modeling the way forward, and encouraging the rest of us to write, report, and create.

To my family and friends for supporting this project, including my brother the Honorable James J. Piampiano for the late night chats on content, and the intrepid Laura Loomer, citizen journalist extraordinaire, for being there to bounce around ideas.

INTRODUCTION

How I Was Stung By The Fake News Media at 16

• • •

I WAS STUNG BY THE fake news media at the tender age of 16.

In the 1970's I won a local beauty pageant that was run by The Gannett Newspapers (the founders of USA Today). In fact, Mrs. Frank Gannett sat in the front row in her finest regalia, well over 90 years old and smiling approvingly as I accepted the win, the roses, and the crown. It was highly publicized because The Gannett Newspapers also owned the local NBC television affiliate. Suffice it to say in Rochester New York and all of Monroe County, a young girl with austere beginnings, emerged from the inner city and was about to be thrust onto the public stage. While accepting the crown, the adjustable band was miss-set and it would not seat securely on my head. In fact, I had to hold it while attempting to take my victory stride in front of 100,000 people on TV with a modicum of aplomb and grace.

So when my first interview with a reporter from the newspaper occurred the next day and she asked me what was something memorable about the win, I innocently told the story of the crown, thinking it would humanize me to thousands of teen girls. Imagine my horror and embarrassment when the article came out on the front page of the Family section and the first line in the piece was "Jane J. Piampiano's first words when she was crowned...Saturday night were "The crown is too small."

"But, but, but," I told my mother, 'I meant it was sliding!" The way they worded the article made me sound like I had a big head or as though I was complaining in the very first moment. I learned quickly that one must be very, very careful about what one says and how one says it when in the presence of the media. For the following year, in my pageant related duties, I greeted ambassadors, I cut ribbons with the mayor, and I represented my city at out of town events. But I was likely the quietest beauty queen on record because I was petrified that I would be mischaracterized. For the next 12 months I was dogged by the local paparazzi whenever I ventured out in my official

Gannett Newspapers, Rochester, NY August 3, 1971, accessed online August 2017

role and in my private life, dodging their intrigue with me.

In that moment I was horrified to think that thousands of girls in Monroe County, who I had hoped would look to me as a role model, would now see me as some teen narcissist with a "big head." Most of all, I was perplexed that the very newspaper that owned the pageant had allowed me to be cast in such a questionable light.

And it was in that moment that I learned never to trust the mainstream media again.

CHAPTER 1

A Sea of New Media

• • •

*"We are now in a sea of new
media to capture the lies."*

ANDREW BREITBART

THE AMERICAN PRESS IS CHANGING right before our eyes. We are witnessing the emergence of a new media with explosive formats, less centralized controls, and an impressive rise of a new generation of journalists – fact seekers, real time witnesses, citizen journalists by the millions, and truth tellers in response to the realization by the American public, and others throughout the world, that the press has been completely corrupted. Nature abhors a vacuum but she also abhors garbage and corruptions. The bubble of deception has burst and these new journalists, Internet personalities, bloggers, and some very savvy every day people, many from other industries and walks of life – are stepping up to replace the mainstream media whose corruption and lies have been exposed. It is happening in response to the traditional mainstream media conglomerates that have operated as an agenda driven, intentionally deceptive, bought and paid for arm of the Institutional Left. This Democratic Media Complex is a cartel of media and political operatives colluding to advance the socialist left ideology of the Democrat Liberal party power machines. The term was coined by American media critic, conservative commentator, and author, Andrew Breitbart after discovering the Democratic party, particularly through the corrupted activities of Bill and Hillary Clinton, was able to control the distribution of news and used it to attempt to destroy the reputations and silence the voices of conservatives of all walks of life. Many believe that through Breitbart's revelations and work with others discussed later in this book, people have woken up to the ruse and deception and are countering this corruption by taking media into their own hands. This effort is exploding through the vehicle of the Internet and its many platforms.

I hosted a locally inspired online TV show in Washington, DC for five years and it opened my eyes to the tremendous power and responsibility of the media. Not formally educated as a journalist, I was surprised at the various reactions to my media role in public. Some people wanted to be near a microphone and others ran for cover. I quickly learned that the American public was beginning to step back and re-evaluate the integrity of the press.

Due to the Internet and the explosion of more and more social media platforms through which information is exchanged, disseminated, or received, we are now subjected to faster and more authentic reporting of world and local events. Information is coming at us at breakneck speeds and in copious amounts that can be overwhelming but we are fortunate to have these new options to get informed in real time and with greater visual accuracy. And in many cases it's lightning speed faster than anything mainstream media outlets can produce. The mainstream media, in their fight to retain their duplicitous reign over what we know and when we know it, appear to be missing the transformation occurring right before their eyes while in some cases they are seeing some of the rise of New Media and reacting with dismissals or attempts to besmirch the existence of it. They are in a relatively rapid dismantling and free-fall as the American people understand more and more everyday about the fraud perpetrated on them by the corporate media complex.

The media philosopher Marshall McLuhan famously said, "the medium is the message" and never has that been more apparent than in the revelations throughout the 2016 presidential campaign. This showed us that this is not Edward R. Murrow's journalism. Murrow was known for honesty and integrity in journalism. Through revelations by Wikileaks as well as emerging New Media icons like Mike Cernovich, James O'Keefe, Alex Jones, Joseph Paul Watson, and Jack Posobiec, just to name a few, we have witnessed the mechanisms by which the traditional mainstream media have not only reported false stories and information but also how they fabricate stories and then report them as "news." This beautiful, precious country requires an honest, truthful objective free media to thrive.

The effort to sustain objective journalism got lost in the sea of greed created by the birth of cable news. Cable news ushered in the tsunami of commercialization: sensationalism = increased ratings = money = corruption and manipulation of the media message. Anyone who caught the TV series Newsroom understands how manipulation of the news is done in backrooms and ends up in the A and B blocks of each cable news show.

Andrew Breitbart, the founder of the New Media phenomenon, emphasized the power of the people...through their use of smart phones and other personal photographic, video, and audio technologies now in everyone's hands. He wrote and spoke publicly that where the mainstream media fails, citizen journalists could replace their attempts to mislead by capturing events and redirecting the narrative. We are blessed to live in a time of such proliferative personal technology. Your camera, your smartphone, your social media-related applications makes you just as powerful in breaking a story as ABC, NBC, FoxNews or the New York Times.

Much of this realization began at the Tax Day Tea Party in 2010 where Breitbart was giving a speech on the National Mall in Washington, DC. He had become a major icon of the Tea Party, a conservative group with grassroots beginnings and a mission to push back on the sliding of America into left wing ideology. In fact he was a speaker at the first Tea Party Convention in early 2010 in Nashville. Initially a liberal writer for Huffington Post he evolved into a conservative after several significant social events created epiphanies on a systematic process by the

Hating Breitbart. Dir. Andrew Marcus. Perf. Andrew Breitbart. Pixel and Verse, 2012. *Netflix.* Web. 15 May 2017.

left to promote the very hateful labels onto conservatives of which they themselves were guilty. He became their crusader, as we shall see in a later chapter and on this evening in DC, he told everyone over and over in the crowd to hold up their cellphones in the air and once they did he brought to everyone's realization that "We have a sea of New Media here to capture the lies!"

The people began to raise their phones and cameras, all kinds of personal devices as Andrew told them, "You are my sources!" People looked at each other in what appeared as a mass epiphany and he explained to them how the Left tried to smear them as racist hatemongers but failed when no one could produce video or audio evidence that any Tea Party member had verbally accosted Rep. Lewis with racial slurs at the Repeal the Bill rally in 2010. A visual emerged of a sea of new media and the crowd cheered as Andrew told them that anyone of them could now break a story!

The idea for this story was born.

Hating Breitbart. Dir. Andrew Marcus. Perf. Andrew Breitbart. Pixel and Verse, 2012. *Netflix*. Web. 15 May 2017.

Since then the victories have been piling up. Early this year, New Media icon, filmmaker, writer, attorney, and author of Gorilla Mindset, Mike Cernovich of Cernovich Media, broke what may turn out to be the biggest story of the year in the revelation that former US Ambassador Susan Rice was instrumental in unmasking classified names in surveillance during the Obama administration relative to President Trump's campaign.

The citizen journalist now has the ability to neutralize attempts by mainstream media to shut down stories that do not fit their narratives. When 25,000 people are tweeting an event with photos, it's virtually impossible for mainstream outlets to shut it down.

Remember - your cell phone is a global megaphone! The Democrat Media Complex, the American press has operated as an arm of the elites and previously they enjoyed the power of controlling information. Now the ability to capture and disseminate information is now in the hands of any person with a cell phone in their hands.

We are now in a Sea of New Media – and it is us.

CHAPTER 2

Many Perfect Storms

● ● ●

Traditional journalism, where reporters deliver information in a balanced and unbiased fashion is rapidly fading into obscurity."

LANCE MORCAN

I WANT TO GIVE YOU my special picture of who I believe to be the founding force of New Media. He had been on the scene and on his mission to destroy the Institutional Left at breakneck speed for years before I heard Andrew Breitbart speak for the first time - just 10 feet away from me at the lectern at the Conservative Political Action

Conference (CPAC) in February of 2012. I was new to DC and a neophyte in the up close and personal world of DC politics, but I was determined to learn more about the Conservative players after perceiving that the Obama first term had gone awry. Part of the collection of catalysts that led up to the Trump movement phenomenon included a Democratic party, under the leadership of Barack Obama, first black President of the United States, a former community organizer and junior senator from Illinois with a murky personal and political history. In retrospect, Obama was just another tool in the Left's armamentarium to transform America from a Constitution-driven republic designed to secure individual liberties to the same failed Marxist dream world of socialism that has failed everywhere else in the world. Breitbart does an excellent job of explaining the roots of this poison in America in Chapter 6 (Breakthrough) in "Righteous Indignation: Excuse Me While I Save the World" and I highly recommend it. In it he chronicles how the disgruntled European Marxist expats embedded the foundation of American communist efforts and it is an incredibly important piece of US history that should be mandatory education for all high school curricula.

At CPAC 2012 I stood in line with other hopefuls for several hours to see one of the movement's sturdiest icons, Newt Gingrich. Surely this was what I expected to be one of the perks of living in Washington, getting up close to the political firebrands and their messages. It seemed more important than ever because of the stakes in the 2012 presidential campaign and the desire to make Obama a one term president after his first term began to reveal his family's communist influence, especially his Muslim anti-American father's activism with which his mother shared, and his communist grandparents both of whom had basically raised him. His policies had become a bell weather for his latent self-afflicted shame of America's exceptionalism and his previous kinship with American communists, Bill Ayers and Bernadette Dorn.

So after missing the cut-off at Gingrich's speech by a couple of people, many of us were herded into a "spill-over" room at the Marriott Wardman Park in the chic little neighborhood of Adams Morgan, a huge convention size hotel in the heart of DC and the host for the 2012 CPAC conference. After watching Newt on a large living room size TV monitor with about 200 others in a dark windowless room, he was quickly escorted out of the building by security and I meandered down to the main ballroom to see what else was happening. The heightened sense of what secret service calls "movement," the activities related to securing a place for a VIP, like increased police/agent presence, the cordoned off of hallways normally free flowing, and the presence of obvious entourages, had passed, yielding a welcomed lull in the action. The room I had previously been shut out from entering to see Gingrich was now wide open, without lines or security; the main ballroom, with two jumbotrons on either side of an expansive stage was nearly empty.

People were slowly ambling in, sitting randomly across the room. Still reeling from my disappointment in missing Newt's speech in person, I charged right up to the front and found a somewhat discreet area in the 2nd row and a bit off to the side. I was told by another attendee that there would be "several lesser known" speakers coming up. I patiently waited, having no idea that what I was about to experience would change my political awareness and my life forever.

First, Conservative British writer, journalist and member of Parliament Daniel Hannan approached the stage and began his discussion. Revolutionary to describe him would be an understatement as he passionately warned us how the United States was the last bastion of freedom on this Earth. At the time he was also serving as the Secretary General of the Alliance of Conservatives and Reformers. He had premonished Europe's decline and told us that Europe was in the grip of a prolonged winter. Most importantly he warned America not to follow in the footsteps of sleeping Europe and not to drop off the cliff of globalist lies. My mind raced with excitement. What wonderful revelations had I gotten myself into? I was reminded of why this is the greatest country on earthy…by a foreign citizen. "If you go down the road toward more government," he implored, "and more regulation and higher taxes, you see how quickly Americans start behaving like Frenchman." Here was a Brit warning Americans to stop the madness toward socialism.

That was my entrée and introduction to the unforgettable, undeniably electric speech by Andrew James Breitbart, liberal college bon vivant turned conservative rabble-rouser. Still digesting the incredibly passionate Hannon talk, I find myself listening to this incredibly charismatic man, about whom I had basically no previous knowledge, but who changed the world as I knew it. Stop and think about what charisma is. Virginia Postrel discusses the differences between charisma and glamour in her work for "The Power of Glamour: Longing and the Art of Visual Persuasion," and Andrew was definitely charismatic. One difference I noted in her description is that with glamour, there is a level once removed, from the object of glamour, a certain indigenous mystique whereas with charisma, there is an approachability factor that accompanies the object person. And Andrew had that warmth, approachability, and self-deprecating persona that made you smile and want to hear more.

Andrew had been a fierce proponent and defender of the Tea Party. By the time I first heard him speak in DC he had already made countless speeches at Tea Party rallies and spent his energies refuting the lies perpetrated by the mainstream media about the integrity and character of the people in the new conservative movement known as the Tea Party. He was not going to allow the Institutional Left to slander and libel the foundation and the intentions of this counter-big government movement. But here he was at CPAC for a speech and he was a already known as a highly controversial person, mostly based upon his production and backing of James O'Keefe and Project Veritas' historical take down of the Association of Community Organizations for Reform Now (ACORN) – a name destined to become the mother of all euphemisms for that organization. ACORN, in reality was ostensibly a non-profit, community support organization with numerous offices across the US, a little Obama jewel that used tax payer funds, as exposed by American hero James O'Keefe, a twenty-something, independent undercover investigator, to fund sex trafficking, prostitution, drug dealing and other heinous corruption. And then there was the Anthony Weiner expose, the fallen New York congressman and pedophile busted repeatedly for trolling young girls on the Internet and sharing pictures of his genitalia. Breitbart knew that Weiner was protected by the Institutional Left, comprised greatly of the mainstream media, but went after him regardless. Weiner married Hillary Clinton sidekick, Huma Abedin whose family was subsequently revealed to be associated with the Muslim Brotherhood. Eventually, Breitbart's efforts to break the story with photographic and social media evidence compelled Weiner to admit his pedophilia publicly. But this was after repeated denials and attempts through collusion with the mainstream media to label Andrew as a "conspiracy theorist."

And then there was the incident where the colluding media assisted the Institutional Left to pin a racial slur that never occurred, on the conservative activists. On March 20, 2010 while walking from the Cannon Building over to the Capitol, three black Congressman, John Lewis, Andre Carson, and Emanuel Cleaver claimed they heard a voice in the crowd call out a racial epithet. This claim, which later was exposed as a race baiting lie, was further backed up by white representative from the State of North Carolina, Heath Shuler who emphatically stated that he heard it. Existing video footage failed to show this. Breitbart was hugely instrumental in calling that lie out by challenging the accusers, some of them prominent government representatives, to prove it with a video from all the videos on citizens (peoples phones) – and they could not. He pledged $100,000 to the United Negro Fund to anyone in the country who could produce footage from the many people there that day. The fact that thousands were there likely recording the event on their phones and none could be found with anyone from the Tea Party shouting a racial slur was one of the first exciting demonstrations of the power of the citizen journalist. Andrew was cultivating all of us to use the power of our own witnessing and video capability. Now anyone could break or verify a story.

Needless to say, Andrew had likely made a few enemies.

His February 11, 2012 speech started out with the living embodiment of the title of one of his works, a must read for any contemporary conservative revolutionary, Righteous Indignation. He was focused and his call to arms to conservatives was very clear – the takedown of Barack Obama. Not for simple political motivations of getting the opposition party replaced with his own, but to right the wrongs, to save the country from the perniciousness of socialism, and to pull apart the lies and rebuild them with the truth. As he once said, "Truth is a trap." Indeed, truth is a trap for those with nefarious intentions around the uniqueness of America.

Groups like Black Lives Matters and "Anti-Fascism" the violent left wing group known by the partial acronym, Antifa were not yet in center focus. But American cities were plagued with a movement called OCCUPY, comprised of the paid or drummed up "shock troops" cultivated by the Left that had "taken over" urban centers by getting crowds of people to set up semi-permanent tent homes on public property. They sprouted the seeds sowed by Alinsky's Rules for Radicals. Their previously effective approach of propagating false narratives and self-righteous banner for tolerance was beginning to be unraveled by people like Breitbart. He said he would stand by anyone in the war against the Institutional Left – it was necessary because this is not your Mom's Democratic party.

We were inspired to hear what we had struggled with for so long – that the mainstream media like MSNBC, CNN, and to some extent FoxNews could no longer portray us as racist, bigoted, white supremacists. We later learned that these were flash mob groups that were essentially hired for pay to "protest" and to this day can be traced back in part to far left wing Anti-American globalist billionaire George Soros enabled by a mainstream media panting in readiness at the Left's beck and call. Contrary to the popular views at the time, agenda driven media was in the early stages of losing its footing in spite of the stronghold optics perpetrated by the Democrat-Media-Complex.

And part of this Democrat-Media-Complex was the American universities. Many were still waking up in the US to the quiet coup of the Democratic Party by Alinsky's community organizers and spawns of the Frankfurt School plan. At CPAC Breitbart explained in those 26 minutes how the American university had (by design) devolved into a tool for the destruction of the American experiment of justice and individual liberty. The university had become, and still is, a crushing force upon free speech. He unpeeled the onion of Obama in response to the question lingering from 2008: Who is this guy Obama, where did he come from, and how did he get on the national stage from a 5 minute stint on Congress and a resume with only two words under Employment... Community Organizer. Barack Obama is a radical. He said it and he told us not to be afraid to call out what we were actually seeing but too politically correct to say among ourselves.

As part of the blowback to the ubiquitous deceptions perpetrated by the Democrat-Media collusion, Wikileaks emerged as a social media force to support the unveiling of corruption in the Democratic National Committee as well as the depth of fraud in the Clinton Foundation. Wikileaks was founded in 2006 by Australian computer programmer, Julian Assange who was an activist in his own right. It functions as a central pivot point or clearinghouse for otherwise privileged documents and communications and its motto is, "We open governments." Possibly inspired to create Wikileaks by Daniel Ellsberg's 1971 release of the Pentagon Papers, Assange built bigger and better placed servers to expand the whistleblowing process for taking down government corruption. There are two major aspects of Wikileaks that separate it from many other such enterprises, its content is not editable by end users and its released content has never been incorrect, fabricated, or unauthenticated.

Wikileaks brought a certain power to social media by opening up confidential messaging of collusions between political operatives, deep state players, and phony establishment congressional representatives. Its work, whether perceived as heroic or traitorous, nonetheless incited an awakening in the American people that spread though social media platforms like Twitter and Facebook like never before.

In 2010 Wikileaks posted nearly a half million confidential documents relating to the wars in Iraq and Afghanistan. President Obama quickly declared them a threat to national security but what quickly became evident was that the threat was to his Presidency and Administration. And during the 2016 Presidential election Wikileaks published over 30,000 of Hillary Clinton's private server that essentially revealed the breadth and depth of corruption and collusion to ensure her nomination and the continuance of the leftward globalist pull of America. These documents, with clear evidentiary materials provided irrefutable revelations that the election was indeed rigged. And it was part of the public realization that the time had come to not only pay attention, but to use social media themselves to push the truth forward and illuminate the corruption in American politics.

Andrew hated the liars and the smear merchants of the Democrat-Media Complex. He charged at Max Blumenthal, son of Clinton crony Richard Blumenthal and half baked journalist for calling James O'Keefe the man who busted the corrupt federal agency ACORN in an epic undercover sting, a racist simply to discredit his legitimate expose. Andrew spelled out the toxicity of innuendo. Sometimes far more dangerous than an out and out lie, innuendo seeds a negative impression and or false belief. Most people can readily identify and reject a blatant lie but innuendo, a statement that indirectly suggests that someone has done something immoral or improper, is damaging to the truth on so many levels and in so many ways. One of the reasons for this is because innuendo, by creating doubt in the minds of people, has extremely long lasting effects. In the form of suggestion it conveys vagueness making it very difficult to defend against. Even as the innuendo forms into a recognizable lie over time, the damage has already been done. Suggestions of nefarious activities or behaviors can damage someone's reputation indelibly and indefinitely. Innuendo is absorbed and wicks into all future considerations and discussions. Disproving or countering rumored statements is difficult precisely because they are vague and incomplete. They also carry an element of sensationalism that attracts people like moths to the light.

No one and no group is more effective at spreading poisonous innuendo then the Democrat Media Complex. In fact if you look carefully at headlines in the mainstream media there is frequent use of innuendo to incite divisiveness and discredit honest journalism and people with real integrity. Seeding doubt in people's minds creates the damage.

Democratic politicians are also masters of the innuendo in attempts to create doubt and foment fear and resistance against the Trump administration and anyone outside the mainstream media. And for a long time it worked as the establishment Republicans sat by in weakness and unexplainable silence. And in another instance, New Media journalists like Mike Cernovich as well as outlets like Business Insider revealed how CNN reporter Andrew Kaczynski ruined the lives of innocent people like Justine Sacco, Sunil Tripathi, and numerous others with a toxic mixture of innuendo and lies. Kaczynski, a well-known press smear merchant's tweet was seen by 80,000 people before authorities corrected his dirty work. Again, the damage was already done.

A Sea of New Media

We now have to sift through more than just lies from the mainstream media...we have to sort through the insidiousness of innuendo thrown irresponsibly into a tweet, an article or a speech like a Molotov cocktail. But I have faith in the awakening of the People and in the unstoppable New Media rising.

All of these factors played a large role in developing the public into the sea of new media that was rising in the United States. This was to become the legacy of Andrew Breitbart's call to action on the national mall that night. And his siren to the American people was, in a sense, an empowerment to take back their free press, which had bee squandered by the mainstream media, and which would become the main mechanism for taking back the country from the globalists and the Institutional Left. He was focused on stopping this by precluding Obama's second term. Some of Breitbart's last public words at that CPAC speech were about vetting Obama in his second term bid and h made the stunning revelation, "I have videos." And in these videos he hinted that we would see the truth from his college days which would underscore Obama's ratcheting up of racial division and class warfare that are central to what hope & change was all about. We sat there with our mouths open and our hearts pounding. We didn't want him to stop. He had stirred something in me and I knew that my life was about to change.

Two weeks later he was dead.

CHAPTER 3

Old Corrupt Media: How Did We Get Here?

• • •

"Let me control the media and I will turn any nation into a herd of pigs."

JOSEPH GOEBBELS

How DID WE GET FROM the ideal of honest objective reporting to multi-billion dollar media conglomerates? That's easy – business people figured out pretty quickly that there is big dollars to be made by marketing the news. And to their credit, the Democrats and leftists did a better job of commandeering almost the entire media complex and embedding it into their system as the propaganda tool. Unfortunately for too long the public has been gullible and "news lazy" which is to say that it became easy to spoon-feed messages to them like hungry baby birds. Many people still see any form of news as truth and reality.

Before we get into what New Media is and what it looks like in everyday life, a review of mainstream media, or what is fast becoming old media, is important as a jumping off point. The current mainstream media grew into a behemoth system of interlocking elite interests during the transition from traditional broadcast networks to purely partisan cable news networks and their departure from authentic information seeking journalism once they realized they could make money. Fact gathering, fact-based reporting of relevant news of events in the world was still discernible in modern journalism until the birth of the cable news networks. Corporate business models replaced facts based reporting in the glamour of television and commercial sponsors with big coffers looking for widespread exposure. Ratings attracted attention and power interests, which were exactly what advertisers were looking for. When high ratings command advertiser money, the sensationalism and controlled narratives that drive those ratings upward then dictate the presentation of news, in terms of content and timing.

News essentially became "theater" bringing the news business to the height of yellow journalism. The term yellow journalism is defined by Dictionary.com (http://www.thefreedictionary.com/yellow+journalism) "as journalism that exploits, distorts, or exaggerates the news to create sensations and attract readers and viewers." Going back to an 1895 cartoon called "The Yellow Kid" that used the bright color yellow as an experiment to attract readers by color and later by the color and sensational, sometimes fabricated content. The Yellow Kid cartoons ran in Joseph Pulitzer and William Randolph Hearst's newspapers in New York.

Wikipedia contributors. "The Yellow Kid." *Wikipedia, The Free Encyclopedia*. Wikipedia, The Free Encyclopedia, 31 Jul. 2017. Web. 3 Aug. 2017

Yellow journalism is the very foundation of today's fake news and its purveyors are now known as the fake news media. It reflects effort son the part of the press to exaggerate and in many cases make up content to get the public's attention, resulting in a devastating disservice to the public.

The "Yellow Kid" cartoons in the late 1800's were designed to attract readers visually when content was not sufficient. Later this would go on to symbolize shameless sensationalism to drive up readership, often with fabricated information.

Those transitions resulted in some serious changes that I believe are only the tip of the iceberg in terms of what is coming. First, the shift to cable news networks created an explosion of news channels: CNN, CNBC, MSNBC, FoxNews, and later networks like One America News Network (OANN). The second element has been a pervasive infiltration of shameless partisanship and this particular aspect of cable news has been incredibly underestimated in terms of its effect on the public. It represents the propensity of the mainstream media to dictate what content we should be getting and it is precisely what the New Media movement is rejecting. But the most egregious part of this tale is the unholy alliance that developed over the decades between the "free press" and one political party in the United States, the Democratic Party to form a cartel driven to mislead and misbrand anything that was not the Democratic Party including the Republicans, the Conservatives, and the Libertarians to name the most prominent. There can be no free press when the entire mainstream media outlets are entirely wedded to one of the two main political parties in the country. This simple fact is something revealed beyond all doubt in the 2016 presidential election.

One of the most negative outcomes of the cable news network proliferation is that it made the industry profoundly inbred. Newscasters themselves becoming the "news" vis-a-vis their faux celebrity and this phenomenon has skyrocketed. News deliverers are now "the news" and they command 7-8 figure salaries with additional power over the content and delivery of the news. No greater breeding ground exists as a springboard to cable news pundit and faux celebrity than an immediate past government position. Unqualified people like Meghan McCain of FoxNews is managing major shows and national political news material with, no offense to art majors, but a Bachelor's degree in art history.

Is anyone else tired of seeing cable news networks creating show after show featuring their own anchors and reporters? This is a very untoward phenomenon proliferating across all of the networks like FoxNews, CNN, FoxBusiness, and MSNBC. It is actually consistent with their propensity to make up news for ratings, which is consistent with yellow journalism. Fabricating, lying, embellishing facts or events with talking heads that are no more informed or expert than the general public is what mainstream media is all about. Former Obama administration like Marie Harf, who was former Acting Spokesperson and Deputy Spokesperson of the State Department, known for her unabashed lying in frequent press conferences in order to mitigate the public ire over the US payment of $1.7 billion to Iran, a state sponsor of terrorism calling for wiping Israel off the face of the earth now unabashedly appearing on FoxNews as a devil's advocate pundit and "FoxNews contributor," is astounding. And similar examples abound on each cable news network under the guise of "fair and balanced." There is nothing fair and balanced about money and ratings. Another aspect that the cable news networks have and are currently overlooking is the indifference of the digital generation to television news programming which is results in self –feeding globalist promoting repetitive, agenda driven news cycles.

When the model is to make up news or repackage information to perpetuate a biased narrative, both of which are designed to spoon feed the public, this creates faux celebrities by featuring their own pundits, reporters, and "contributors" over and over again until their recognition factor begins to create and justify a form of celebrity. This is not by accident. Cable news is all about faux branding to generate ad sponsors. By creating faux celebrities with their own people, they are creating a fake branding for as long as they can milk it. And along with the celebrity created around some of these mundane characters, there is an insidious suggestion that they are experts upon that which they opine and a simple tracing of their education and credentials reveals no such thing.

Some examples include Fox's "The Five," comprised of five people who are on a number of other shows throughout the week interviewing politicians, public persons of interest, and yes, you get it, each other! And then there is "The Specialists," another group of already employed Fox pundits who, for the life of me, do not reflect any element of specialization in anything. The "poster child" show for this could be "Media Buzz" hosted by a liberal media bon vivant Howie Kurtz who regurgitates the week in the eyes of the media...that is more about repackaging the week's fake news media garbage. And this is a show? CNN tries to disguise the inbreeding of a show like "State of the Union" biweekly show with the tag, "Interviews with top newsmakers on politics and policy—covering Washington, the country and the world—delivering a unique and comprehensive look at what matters most to you." Nothing could be further from the truth. But CNN as well as the other mainstream establishment elite media are not in the business of fact gathering or truth telling. Yes, a politician or outside expert is occasionally sprinkled into the mix but the main players are the same faces, under CNN contracts, that are force fed to viewers over and over. Self-avowed communist and now whitewashed "moderate" Van Jones, Republican Congresswoman Marsha Blackburn, former politician liberal Democrat Jason Kander are recycled day after day and the list goes on.

To deepen the inbreeding and create a façade that there are differences between establishment Republicans and the Institutional Left, both FoxNews and the more generally liberal outlets such as CNN and MSNBC have these self-declared experts" and pundits from the opposing political side. And they the trade visibility of theses faux celebrities across networks. After watching former State Department spokesperson Marie Harf lie her entire way through the Iran Deal including the $1.7 billion gift to Iran courtesy of Barack Obama, we now see her as a "contributor" on FoxNews.

All of this inbreeding and further development and entrenchment of the Democrat-Media Complex, the press arm of the Institutional Left is about to explode by virtue of the Murdoch boys' takeover of FoxNews. If you thought that FoxNews was the last bastion of a modicum of New Right friendly, America First agenda, or non-establishment conservatism, think again.

There's a reason why lines for Tomi Lahren were around the block of the convention center or that someone like Charlie Kirk commanded over thousands at the 2017 Politicon. They didn't show up to see mainstream media lying junk press personalities like Jake Tapper. People are hungry for a new and different media deliverable.

The result in just 2017 alone has been the emergence of New Media personalities and forces, to create brand new media outlets and distribution control houses, including New Right networks like Alex Jones's INFOWARS, Mike Cernovich's Cerno News / Cernovich Media, and Ken

McClenton's DC-based The Exceptional Conservative Network.

The fake news media comprises all mainstream media and most of their counter-parts. We are in an era of disruption and unmasking of the corruption of the establishment media and the deception perpetrated on the American public by a press that was granted a public trust under the First Amendment to the Constitution of the United States.

It is time for more than just the presence of New Media calling out the fake news mainstream media on their fake stories, disregard for real news, and numerous collusions with the Institutional Left. My prediction is that the American press is in a state of deconstruction and destruction whether they see it or not. The deconstruction is happening by virtue of the fake, false, and biased agenda-driven operations in the MSM being exposed everyday by New Media warriors discussed in Chapter 5. This deconstruction and the destruction will be neither smooth nor will it be comfortable for the MSM. A catharsis in the form of debunking must precede the full establishment of the New Media. But make no mistake, it is happening and there is no way to stop it now.

CHAPTER 4

New Media: What Does It Look Like?

• • •

*"New media is like a megaphone; it amplifies
your ability to reach more people."*

MARK BATTERSON

WHAT IS NEW MEDIA? It is too simple to describe something that is a culture changer merely as everyone with a smart phone. First, it appears in many forms we categorize under the Social Media rubric such as Twitter, Gab, Facebook, Snapchat, and blogs. But the forms are vehicles for New Media; they are not the New Media. And for now, they live primarily on the Internet. And this is where its vulnerability lies – in a sphere of traffic that, pardon my paranoia, can be shut down vis-à-vis censoring (or disappearing from public use) at any time. But New Media has one common powerful thread, one tremendous super-power that eludes the mainstream media's grasp: Sharing. The power of sharing, outside of the mainstream media is precisely how it will lose its power and grasp. New Media by its very nature has the ability to create exponential exposure, across huge geographic lines, at breakneck speed.

There are countless examples. Several days after Hillary Clinton collapsed after cutting short her visit to the 9/11 tribute in New York City in 2016, The Hill published an article entitled, "How a smartphone camera changed the discussion on Clinton's health" that detailed the story of how Zdenek Gazda, a New Jersey resident and former firefighter/immigrant from the Czech Republic filmed Clinton's now famous get away from the 9/11 tribute in New York City with stunning footage showing her collapse as she was being helped into her van. After he uploaded the footage to his Twitter feed, it was ultimately viewed by over 11 million people and re-tweeted over 100,000 times. This demonstrated that one person, with a cell phone, could change the course of history and even more importantly, that in a sea of new media,

A Sea of New Media

the mainstream media no longer controls the narrative. The article concludes with a statement that is at the core of this book: "Citizen journalism. It's changing the way campaigns and the media do business as usual forever."

For months leading up to this event, filmmaker and author Mike Cernovich had been pushing memes to alert the public to Clinton's questionable health and taking criticism from the press and others that he was a "conspiracy theorist" a term used by the Left to shut down alternative explanations for events. But Hillary Clinton had been displaying bizarre episodes like speech stopping intractable coughing spells and a startling head jerking, seizure-like episode one day while she was being

interviewed by a gaggle of mainstream reporters. Still none of this stuck because the mainstream media was in full deflection and cover up mode.

This expose of Clinton collapsing and being dragged into the SUV was a shocking revelation that there was a serious medical issue going on and that a media cover up was in effect.

Events like this are clearly signals that media is changing. With New Media and all of us chronicling events in real time we overcome the slow drag of mainstream media. They take extra time to gather information and to sort it into their desired news block for dissemination. A news block is a form of triaging or prioritizing news topics. The blocks are buckets of news stories from A to D or E. They run "A" block first so anything in the A block is high priority and you should pay attention to that to determine what <u>they </u>determine (according to often hidden agendas) you should be focusing on. They also take extra time to conduct content or appearance "spin" on what is presented. With New Media, there little time for hidden agenda spinning – people capture events in real time either in still photo or by video and then upload them into media forums like Twitter or Facebook. Even more exciting is the ability to live broadcast right where you are by using such applications as Periscope. And my prediction is that there will be many more platforms coming with increasing capabilities to enable us.

Another example is my own revelation where I was recently deplaning in Raleigh Durham to connect to another flight. Within minutes of reaching my connecting gate, system-wide alarms went off and an announcement looped over and over telling us that there was an "undetermined emergency" in Terminal 2 and all must evacuate. My first instinct was to assess the situation and hold to observe until the airport authorities arrived to guide this emergency evacuation. The announcement changed to indicate that there was a real fire emergency going on in Terminal 2 and reiterated that evacuation direction.

I started to see chaos in the situation and I thought of Breitbart's words and I thought about all of the New Media warriors I will discuss in Chapter 5, The Media Equalizers, like Mike Cernovich of Cernovich Media and Jack Posobiec who generously share their vision and expertise and encourage the rest of us to film it, tweet it, and get the story out there. There are growing numbers of social media icons that are prime examples and guide posts for everyone, the average everyday person, to capture a situation or story in real time and get it into the public sphere. This is an incredibly powerful new phenomenon in the 21st century. Other well known people like Stefan Molyneux, Joseph Paul Watson, former actor and conservative James Woods, Alex Jones, and former Nixon and Trump advisor / political provocateur, Roger Stone are propelling the anti-left false narrative efforts exponentially in collaboration with the general public. This is an unprecedented relationship between any form of media and the public.

So I pulled out my phone and captured the chaos. The next day I was interviewed by ABC11 out of Raleigh Durham to describe the situation. The critical point of this story, which would have been lost without my report on Twitter, was that the Raleigh Durham airport did not

appear prepared for emergency readiness and had this been a real fire or if there had been an active shooter in the terminal people could have died as they scrambled on their own to flee at the behest of the overhead announcement. The other issue is that without the footage from witnesses on the scene, the airport's absence of emergency readiness could have been hidden from the public.

In summary, this citizen's report may save lives next time around.

While many people agreed that Barack Obama introduced the use of social media to the campaign process and Andrew Breitbart, in his mission to defeat the Institutional Left, laid the foundation for the emergence of the New Media, it was Donald Trump who was first to use social media to achieve direct communication with his audience. Obama used social media but in retrospect, he did not cross the boundaries of the mainstream media. How could he? They were his most powerful allies and he knew that they would grease the wheels for his slide into the White House. There was no need nor would it have been useful for him to go around the traditional press. The magic of bypassing the press was bubbling quietly under the surface, waiting for the appropriate time to surface eight years later.

The power of using social media in a new way, a way that marginalized the traditional press was using it to bypass the gatekeeping effect of the mainstream media and speak directly to the people. Donald Trump had a clear and unshakeable grasp of this and used it almost like a brand. We watched his Twitter following grow from 1 million all the way through and beyond his election to the roughly 35 million followers. His Facebook following tops 23 million at the time of this publication and if you include the 7 million Instagram followers, you have an impressive reach of 65 million people. Even if some of them overlap, it still represents hundreds of millions of communication impressions across the US and the world.

New Media is Memetic

The mainstream media was historically the gatekeeper of all things news by parsing out what and when we were given the news. Donald Trump the candidate readily saw the power of bypassing the press and taking back the country from the elites using memes to effectuate psychological persuasion. Memes are concepts that have received tremendous attention in contemporary communications and they should because they can make or break a reputation, a policy hoping for support, or a candidate making a point. Memes represent the next level evolution of an expression. Out dated expressions like "penny wise and pound foolish" readily communicated a paragraph worth of communication in just 5 words. Once defined for you, you can understand that it represents someone who is overly careful with smaller amounts of money while ironically having no compunction to spend large amounts of money at other times. Worrying about the smaller amounts is not helpful if one is indiscriminate with larger amounts. In the same way, memes have come to communicate a thought, a feeling, a warning, a reality of life in a visual, sometimes in motion, expression. Memes can also be communicated as a verbal expression. Certain fads or catchphrases gain momentum or "go viral" by virtue of society relating to them and then propelling them out into the Internet. Social media provides the freedom for the everyday person to propagate information at exponential rates known as "going viral" or spreading rapidly. Coined by English evolutionary biologist and philosopher, Richard Dawkins, the term emanates from his use of the Greek word *mimeme*, meaning, "that which is replicated." It can also be a pervasive thought or thought pattern.

New media is memetic and it us important to note that memetic communication either by concrete visual (both still photo and moving visuals) or by creating a powerful mind visual, it not a not literal translation. It is a powerful tool to move the culture in a particular direction. Many New Media personalities and leaders believe that those on the political right are much more adept at meme making than those on the left because the left has been so entrenched in more traditional forms of persuasion that could be relied upon.

Visionary entrepreneur and political thought contributor, Jeff Giesea has written a great deal about memetic warfare in NATO related publications and other forums frames it this way:

"But for many of us in the social media world, it seems obvious that more aggressive communication tactics and broader warfare through trolling and memes is a necessary, inexpensive, and easy way to help destroy the appeal and morale of our common enemies." (Defence Strategic Communications: The official journal of the NATO Strategic Communications Centre of Excellence, Volume 1, Issue 1, Winter 2015; ISSN 2500-9478)

To see an excellent example of a meme war against the mainstream media, known now in the post 2016 election period as part of a cartel within the Democrat-Media Complex that is nothing more than agenda driven purveyors of concocted narratives disguised as "news," Trump propagated a gif or repeated cycle of a video clip showing past real footage of him spoofing a punch at someone in a WWE fight with the victim's head and face superimposed with the CNN logo. Having used the verbal meme of "you are fake news" repetitively in presidential press conferences at well known CNN journalists whose work had been unveiled as lies and false narratives, the video meme of Trump punching the CNN headed figure was profoundly effective at saying, I will punch fake news out. As we've said previously, memes are not literal – they are visual props of persuasion.

Those who readily grasp the concept of symbolic persuasion have picked up Memetic strategy in our culture. This next generation of New Media creators are also able to translate their technical, video, and audio hyper skills into this artistic and culturally provocative form of communication and persuasion. Mainstream media and the key players on the left are stuck in the old stale forms of communication: the use of literal, concrete symbols and words that are growing increasingly ineffectual. This is because the forms are overused, antiquated for the most part. But the most critical factor attributable to the failure of old forms of media communication is their heavy use in creating false propaganda types of information and dissemination resulting in a national and international loss of credibility. The New Right/New Media movement seized on this stale and false communication history of the mainstream media and coupled it with the leftist political fatigue of the American people after years of establishment Republican corruption particularly in collusion with the Democratic party's huge leftist pull on American politics and policy.

Donald Trump's use of labeling resulted in prolific Internet memes. Not surprisingly, he demonstrated mastery of the ability to exercise a sharp tongue in the primary debates. One of the most impressive approaches to shutting down his opponents was the use of functional sarcasm and body language that was effectively offensive. He labeled his opponents with a visual sumptuousness that fed into incredibly tenacious memes. The use of the tactic was dubbed "linguistic kill shot" by Dilbert creator Scott Adams. Who could forget "Crooked Hillary" (Hillary Clinton), "Lyin' Ted" (Ted Cruz), or "Low Energy Jeb" (Jeb Bush)? The value in a linguistic kill shot is its reflection of what people already believe or see in the person *and* their inability to defend against it. Attempts to defend reduce the target to a defensive position that renders them weak.

So powerful was the use of this toxic labeling that Trump was, in part, able to knock out seventeen opponents. Memetic strategy and warfare is where the future lies in persuasive communication and its sophistication is unparalleled in the history of media. The use of memes is tantamount to persuading others to think in a certain direction by advancing the argument beyond verbal construct more toward a visual depiction of a theme as it communicates the argument.

Memes began to saturate the Internet, resulting in a pernicious branding of his rivals. In a 2015 Forbes article the author explains that the use of linguistic kill

"Low Energy Jeb" Notice the look of boredom in attendees at a Jeb Bush rally, including one particular woman catching a seated nap. From The Daily Caller, "Trump's 10 Best Insults" accessed online September 7, 2017; http://dailycaller.com/2015/12/30/counting-down-the-10-best-trump-insults-of-2015/

shots requires the use of "an engineered set of words that changes or ends an argument decisively." According to Scott Adams, when Trump ascribes these labels to an opponent, "he's imprinting a label you already feel about these people. They're not random insults, but linguistic kill shots that you can never get out of your mind." (Accessed online September 7, 2017 https://www.forbes.com/sites/ralphbenko/2015/11/28/donald-trump-political-mass-hypnotist/#62a7837752fd)

So after numerous other reports cited Hillary Clinton's suspicious dealings as Secretary of State and

"Crooked Hillary" Bombshell Report: Crooked Hillary Took $100 Million From Middle East Regimes: "Massive Conflicts Of Interest" https://shar.es/1TEcMk, Accessed Online 3 Aug 2017

regarding the Clinton Foundation, it was a short leap to Trump's shot of "Crooked Hillary," and made for great chanting at his rallies.

The election was the perfect time and setting for the nascent phenomenon of bypassing of the main press because people had begun to realize that the press had become complicit with the Institutional Left, comprised of the Democratic National Party, the Liberal factions of government, and the grip of the wealthiest cable owners and influential advertisers the country had ever seen. As the candidate Donald Trump violated all conventions and pushed his opinions and labels of his opponents through to the American people, we saw their awakening spread exponentially across the country. Despite the unrelenting criticism by countless members of the mainstream media and the establishment Republicans in the party, Trump sensed the receptivity of the people to his direct and audacious messaging. In a word, they loved it. And the more the mainstream media balked, the more the people supported it.

It was clear that by using all of their efforts to destroy Trump's candidacy, the mainstream media was inadvertently laying the foundation for Trump to go around them directly to the people. And it wasn't just Trump using that to his advantage, millions of Americans were realizing that they too could directly communicate messaging and protesting directly to others on the Internet. The mainstream media was rapidly losing its gatekeeping power and I believe that it is this central phenomenon that will be most instrumental in the full and final destruction of the mainstream press.

New media investigative reporter Laura Loomer has dedicated her career to teaching average citizens how to capture and break a story in real time. Her perspective and approach will be discussed in Chapter 5, Who Are the New Media Warriors but her idea is incredibly organic. Here is an example of what she is trying to propagate in this new media era.

In the summer of 2017 I was invited to a press conference by Maria Espinoza, founder of The Remembrance Project – an anti-illegal immigration non-profit organization in the United States based in Houston and Washington, DC. According to their website the organization is dedicated to "Educating and raising awareness of the epidemic of preventable killings of Americans by individuals who should not have been in the country in the first place." (http://theremembranceproject.org/).

The mainstream media was invited but when I arrived there were only two reporters in the room and the presser was about to begin. As a novice and observer of new media at the time, I was only there to support the organizers but when I saw the low turnout it struck me that I could capture the presser on a live stream platform like Periscope so I asked for their permission and I live streamed the 15 minute event. Fortunately and because of the exponential power of social media, a social media personality with hundreds of thousands of followers shared my live broadcast and over 600 viewers joined. At the end I informed the organizers that it was as valuable as if 600+ people were in that room attending their pressers. They were astounded at the realization but more importantly, they immediately grasped in that moment, the sheer power and reach of new media through social media.

As we've seen, social media is only one part of the phenomenon of New Media. Social media is the main vehicle for advancing New Media and it is more about the way people and the type of people that are using social media to shape the proliferation and the power effects of New Media. There are two key players in the New Media phenomenon and both played a major role in the 2016 election.

And as for the 2016 Presidential election? It was more than "the social media election" – it turned out to be the New Media election through social media.

CHAPTER 5

The Equalizers: Who are the New Media Warriors?

• • •

"We're part activists, part journalists, part provocateurs"

MIKE CERNOVICH

LET'S TAKE A CLOSER LOOK at who comprises the New Media. One of the most interesting aspects of the rising visibility of those who are moving the New Right messaging through social media is that many of them emerged from the general public with little to no media experience and some of them had made a name for themselves in other, politically unrelated industries or areas. Some of them were lured into this New Media world such as blogging, podcasts, website columns and other conduits of communication such as those through applications or platforms. Some of these platforms included the usual forms like Twitter and Facebook. Others sprung up out of the organic demand for unique features in communication such as Periscope for live streaming an event or discussion. The explosion of this citizen-to-citizen communication was never more apparent than during the 2016 US Presidential election. In describing the people of the New Right who are also leading the New Media revolution, New Right leader and social media maven, Mike Cernovich summed it up perfectly in a tweeted response to an article that asked the question, What's the think tank of the Trump era?" with "We're part activists, part journalists, part provocateurs." (Accessed September 7, 2017; https://twitter.com/Cernovich/status/902372442840195072).

Cernovich's description is ahead of the curve because the old labels and descriptors no longer suffice. In each generation there is an influx of new thinking and an infusion of sophistication from previous cultural iterations. In taking a closer look at the predominant characteristics of the New Media figures that have been observed you might find the following:

1. A profound interest in reporting authentic live as they happen events and information;
2. A shared mission to decimate the mainstream media as an arm of the institutional Left
3. Dedication to correcting false news stories, narratives and harmful innuendo about people and institutions whether on the Right or anywhere in between on the political spectrum
4. Use of transferable skills from previous or other experiences to New Media actions and efforts
5. A devotion to moving the USA back from the left leaning socialist & globalist policies of the 44th Administration under Barack Obama.
6. Restoration of a glorified America, of God, Family and Country
7. An inclusive, socially tolerant open door to the movement requiring nothing more than love of country and the absence of racism and fascism.

Citizen Journalism

Let's take a closer look at Citizen Journalism. Simply, a citizen journalist is someone with a cell phone, some apps, and the Internet who investigates situations and people to capture the truth. It can take the form of blogs, podcasts, live streams, written articles, short stories, video reports, audio recordings all of which is integrated through the Internet's web based innovations. There are two general categories of citizen journalists: The Semi-Independent and Independent. Semi-Independent Citizen Journalists are integrated to various degrees, with a professional news organization or web-based outlet. This integrated role often operates on a contributing basis to provide content to news stories. They act as typical on the ground reporters and through website feeds they will present the information on past or breaking stories. This is similar to a traditional journalist in mainstream media who provides source content. The Independent Citizen Journalist by contrast is a citizen who works fully detached from professional news outlets. Rather than function in a contributory role, this type of CJ is the writer, reporter, and tech savvy delivery news and communications expert. These people can be ordinary people who happen upon an event unfolding and capture it with their cell phone camera or someone with significant media training and experience in the field who is building a brand and a career in this new focused journalism. Their reports can be in any number of the formats like screen content or fixed online print. In some of the online video reports you will also see them shift into political punditry and other forms of commentary. There is a great deal of "in the moment" journalism that was historically more prevalently seen in wartime embeds on behalf of major news outlets.

The most exciting aspect of the Independent Citizen Journalist in my view is that the complete control over what and how a story is covered. Mainstream media outlets had/have the same degree of control and choice however the current public sentiment is that they abused the public trust by fabricating stories, pushing biased agendas without disclosure, and manipulating the reports of news events to support the Democratic and Liberal party agendas.

Mainstream media has let the American people down and nature abhors a vacuum. Conservative media giant Andrew Breitbart called to action crowds of people in his speeches by reminding them that they were a sea of media to capture the lies of politicians and the complicit mainstream media. Never has there been a better time for the explosion of citizen journalism.

Citizen journalism is dependent on the exponential reach of the Internet and social media platforms like Twitter, Facebook, Instagram and many others. In fact the Internet is the lifeblood of the New Media – the growing phenomenon of real time, digitally transmitted, and highly interactive communications. Apart from the fact that most social media platforms are run by liberal and socialist activists, a fair amount of conservative communication gets through.

If it weren't for citizen journalism we may never have known that something was terribly wrong with Hillary Clinton who was captured collapsing as her crew tossed her into an SUV the morning of the 2016 tribute to the 911 terrorist attack in Manhattan by passerby Zdenek Gazda, as we saw in the previous Chapter. I myself had the presence of mind to capture a near disaster at the Raleigh Durham airport when an emergency evacuation was called and botched by a lack of an organized process. The next day I was interviewed by the local ABC affiliate and my story went national.

These are very exciting times. Everything is changing and the old labels no longer work. People all over the world, many with little or non-traditional journalism education are becoming journalists: gatherers and disseminators of events in our world. And others with some traditional credentials are fast learning that they must adapt to a "real time" news product. Wikipedia's description of Citizen Journalism captures all of the elements:

> "The concept of **citizen journalism** (also known as public, participatory, democratic, guerrilla or street **journalism**) is based upon public citizens playing an active role in the process of collecting, reporting, analyzing, and disseminating news and information."

There are plenty of examples of citizen journalism and the work can be found on social media. New York City based Laura Loomer, an independent investigative journalist is rising among the ranks of the New Media. She has a passion for what she calls, guerilla journalism and it shows. Guerilla journalism is an in-your-face type of investigative journalism, mostly done by non-mainstream reporters that a number of new media icons are bringing back. Loomer, a former under cover investigator for James O'Keefe's Project Veritas, Loomer gained even more notoriety for storming the stage in New York's Shakespeare in the Park's rendition of Julius Caesar, a play depicting the assassination of a President Trump

facsimile. And this landed her a prime time interview by Sean Hannity on Fox News.

Later this year Loomer waited in line for 7 hours in New York at a Hillary Clinton book signing event for "What Happened," in order to ask Hillary on camera, what happened to her 33,000 emails, the Clinton Foundation donations for the people of Haiti, the Americans in Benghazi and much more. As secret service was escorting her out, she still had the presence of mind to ask Huma Abedin, Hillary's sidekick, why she was still married to a pedophile, convicted sexual predator Anthony Weiner. Again, she was invited to appear on Fox Business for another nationally televised interview to explain why she confronted Clinton and why it was important in terms of keeping her accountable.

A Sea of New Media

In some ways taking the power out of the hands of the mainstream media and watching real everyday citizens replace the stories, reports and ultimately the narrative for persuasion, represents the evolution of a push back from nanny state mentality back to a free and independent press. This puts the onus back on the general citizenry to validate, fact check and make a summary judgment on the information and news coming at them. To be honest, many in society have abdicated that responsibility and most are now awoke to the fact that we were lied to, manipulated for political reasons, and deceived by corruption that emanated from a cartel that formed: The Democrat Media Complex.

While some in the general public and more in the mainstream media mock these New Media warriors, they continue to build followings, garner attention, and most importantly, create change. Even if the story is mangled by the mainstream press, the mere visibility that results can have staggering effects on public policy, behavior of politicians, and mainstream media mischief. Mike Cernovich made such an issue (cleverly) about Hillary Clinton's suspicious health problems during the 2016 campaign that it landed him an interview on CBS' 60 Minutes where he trapped interview Scott Pelley into admitting that he had taken the word of the Clinton campaign to debunk her health problems in spite of her incessant hacking and photos showing her need for assistance to climb stairs.

The mainstream media concocted propaganda for nefarious purposes and this led to the citizenry that is much more skeptical of their agendas and everything they broadcast. As I alluded to above, in New Media we will need to vet sources and wait for story details to play out but it's still much better than the lying Democrat-Media Complex we had. One could make the argument that it is even more important to stay alert, be informed, play a part, and conduct background research before accepting any news offering as valid. Mainstream media in television, radio and print are dying. If you are in a situation that is breaking news or could be used to tell the truth about an event, seize the chance to capture it on your cell phone or better yet, use applications on your phone like Periscope to live stream what you are seeing. And when former naval intelligence officer and New Media icon Jack Posobiec was linked to a massive computer hack of the French president-elect's campaign emails and accused of trying to sabotage Emmanuel Macron just days before that election, even the French government responded, catapulting it into an international event. Social media is real media.

To the naysayers and critics sitting home on social media tearing down those who are creating this breaking phenomenon, I say adapt or die.

CHAPTER 6

Suicide Watch: Implosion of the Mainstream Media

• • •

"Never interrupt your enemies when they're making mistakes"

NAPOLEON BONAPARTE

In the course of the Internet > social media platform > New Media explosion, the mainstream media reactions and actions have been astoundingly self-defeating, self-revealing, and self-destructive. The Institutional Left Cartel, comprised of the Democrat National Committee, the Military Industrial Complex, and their literal partners in crime, the mainstream media, have been exposed for colluding in the propagation of outright lies, the destruction of lives and reputations, and fraud perpetrated on the American public. They have become so drunk on their own power and why not? For 25 years they have been building this power on the basis of a growing corporate media oligarchy that turned a once free press into a stifled, agenda driven cartel, paying very little attention to the groundswell of public disdain for them.

A LOOK AT THE EFFECTS these New Media warriors and the enlightened media-savvy public have had on the evolution of "news" is in itself newsworthy. While mainstream media and the establishment on both political aisles vociferously proclaim that social media is but a trite side show beholden to their powerful news dissemination machines, it is becoming more and more clear that social media has opened a whole new door to inform and influence.

Initially the mainstream media outlets ignored social media comments, reports, and information dissemination efforts. Like an annoying gnat in their midst, many outlets just ignored the existence of the New Media's burgeoning growth but as their visibility and power have grown, they are finding it increasingly challenging to ignore them. Part of the evolution of New Media includes the development of independent news outlets or organization. Breitbart News, started by Andrew Breitbart and Steve Bannon, was one of the first to appear. Alex Jones who burst onto the media scene with fervor and fanfare, started a nationally known online news show called InfoWars and a massive website for international news. The mainstream media is being outflanked and just like the Democratic party's reaction to losing all in the 2016 election, their reaction is to stomp their feet, continue the fake news agenda driven narratives that lost them the public's confidence, and forge ahead in their elite cocoon.

Much of the implosion of the mainstream media is self-inflicted. Shifts are occurring but most are not going to serve the mainstream media, including the cable news networks very well. Fox News that had by reputation and under the leadership of Roger Ailes, had been known as the conservative voice in cable news began to show cracks during the 2016 Presidential election. Their familiar motto "Fair and Balanced" had almost mysterious disappeared from website banners and nightly show host mantras. Most of this can be traced to a shift in management. Fox News could easily be considered a dying, from a conservative perspective, entity.

FoxNews fired a number of conservative male show hosts under a whirlwind of sexual harassment charges fueled by long time female colleagues. Suspicious at the outset, many were settled so we may never know the truth, but the key accomplishment of this chaos was that most of the conservative thinkers were ousted. Take the case of conservative show host and Fox contributor Eric Bolling.

He suffered the fate of predecessors Roger Ailes and Bill O'Reilly—both stalwart and successful conservatives by a number of metrics. People on social media began to ask, could it be a coincidence or act of fate that FoxNews is bleeding conservatives?

Many admitted feeling a gnawing sense over the preceding months that the tone and tenor of the stories and pundit views were sliding over to the Left. They were noticing how the once fiercely conservative and devotedly patriotic Stuart Varney increasingly began to snipe at some of the President's actions or statements that he once found powerful and influential?

The first clue was the disappearance of their "Fair and Balanced" motto which was replaced with "Most Watched, Most Trusted." What we were actually witnessing was a corporate self-cannibalism: the process by which an entity destroys itself by killing off its own inner parts. In self-cannibalism the entity eventually eats itself to death. It's a business risk destined to fail. The last thing the marketplace needs is another progressive propaganda mule for the institutional Left but let them continue on as this will likely be a foundation for their implosion and hopefully demise.

Meet the Murdoch boys James and Lachlan, ultra liberal sons of 21st Century Fox owner Rupert Murdoch. This little alt-Left billionaire cabal is executing an intentional systematic conversion of FoxNews from the once well-known conservative news outlet into yet another cog in the wheel of the Democratic Media Complex—the Institutional Left of which both James and Lachlan Murdoch so revere. The New York Times reported in April 2017 that the brothers replaced their father's Republican lobbying chief with a Democratic one.

Let's take a look at how this all began. Papa Murdoch, although a liberal himself, was willing to compromise culture for the financial rewards of running a conservative news network. Once he handed over the reins to his two ultra liberal sons, they declared war on conservative news and decided they were going to transform FoxNews into something with more "corporate responsibility." Add to that the climate change hysteria of James and his wife Kathryn who works for the Clinton Climate Initiative and openly expresses her disdain for the President on social media, and you have a recipe for Barack Obama's version of hope and change. The Murdochs are Clinton

Foundation hacks determined to turn FoxNews into a progressive propaganda machine through a covert and clever Alinsky tactic of smearing the surviving conservatives into elimination.

Getting Ailes and O'Reilly out were only the first steps in removing huge barriers and expediting this nefarious plan. Their ousters expedited the transformation. And because it works so well, they utilized the same approach in the attempt to take down Sean Hannity and Charles Payne. While Sean fought back and won a temporary reprieve and Charles is coming back, Bolling was not so lucky. And other previously noted conservatives like Varney and Jesse Watters, who of late might be spending too much time with establishment hacks like Dana

Perino, "diluted" conservatives like Gregg Guttfeld, and left wing personalities like Juan Williams, are beginning to show signs of moving left. Are they being pressured from the top to go along to get along? Selling out? Nevertheless, it appears as thought a pattern has begun. The NY Times described the process the Murdochs are using to take down their internal conservatives, (something Andrew Breitbart called destruction by innuendo):

> "It doesn't matter if O'Reilly or Ailes did or didn't do the things they are accused of—no trial has occurred, no evidence has been released, no investigators' conclusions shared—their real guilt is that people believe they could have."

The Murdoch boys have a formula:

Dr Jane Ruby 🏴 💜 @DrJan... ·10h ˅
The Murdoch strategy:

Set up the smear
Let target struggle
Hire Paul Weiss to "investigate"
Feign corporate powerlessness

Fire target

Regardless of the theories behind how it is happening, there is a rising liberal creep at Fox. It is characterized by a majority of left-wing pundits, deep state and Obama holdovers in high profile positions, a pervasive invasion by establishment narrative. We are witnessing people like Marie Harf, former state department spokesperson who lied daily about the Iran deal, now employed as a regular Fox news contributor It is also obvious that the conservative tone and tenor among the resident conservative anchors and pundits has now been noticeably diluted by equivocation and agreement with their left-wing on air colleagues.

If social media is any indication, and I believe it is, the public isn't buying it. In fact the more Fox attempts to make it look like this is just corporate housecleaning, the angrier people get. The best way to stick it to them is to stop watching and/or cancel cable. It's the only language they understand. In the meantime, the vacuum is now open for real conservative news networks to spring up especially through the New Media movement.

CHAPTER 7

New Media Rising: It's About The Movement

● ● ●

"You 'conservative' pundits still don't get it. Trump isn't our candidate. He's our murder weapon and the GOP is our victim. We good now?"

BLOGOSPHERE TALKING HEAD

IT WAS NEVER ABOUT ONE man. It is about the movement. And the movement is not just political; it's cultural, social, and psychological.

A large part of the movement is about bringing America back from the left wing socialist ideology bill of goods that was what Obama's "Hope and Change" were really all about. Pulling the country back from the globalist New

World Order agenda to transform this big beautiful free country into a strangled subdued faux utopia – for the few and the rich. In the summer of 2017 I flew to Los Angeles to attend a seminar in Mindset by the author, attorney, national security reporter, and filmmaker Mike Cernovich. In a totally non-politically focused 4 hours, one of the most impressive things we discussed was how another's criticism of us is basically a projection of them onto you. Those who remember Psychology 101 in college may recall that this is one of the primary defense mechanisms in the human mind. And it is important to remember that human defense mechanisms are very powerfully driven. They are tied into the first primal drive – the will to survive. This is an important concept when being attacked by any opponent.

For decades the Democrat Media Complex has successfully tainted and labeled anyone on the Right, conservative side, or just any free thinking individual or group with untruths and smears. And for the most part, much of it has stuck. Take for example the successful campaign by the Democrats to paint the Republican party as the party of racism, bigotry, and white supremacy ideology like that of the KKK. As a group nothing could be further from the truth but until recently with the Wikileaks revelations of DNC operatives like Debbie Wasserman Shultz, who disparaged blacks and other minorities in their communication, these messages revealed whom the real racists and bigots are.

Everything has changed. The general base of President Trump is somewhat separated into two mindsets. I intentionally do not use the words divided or split because while we have two different views relative to Trump vs. the movement to bring the country back, we share the same devotion to that America First, smaller government, establishment purged vision of our country. This difference in view is important because New Media plays a significant role in the communications that have formed these perspectives. But first let's look at the two mindsets. I will refrain from using labels, either those currently circulating or any by my own wordsmith tendency. One mindset is that the election to take the United States back from the far left yanking it has endured over the last 50 years to be realistic, was heralded by Donald Trump who in his declaration speech in 2015 articulated the primary issues tanking our country. Finally someone who was not in the establishment elite was saying what had to be said. That the socialist takedown begun by the misanthropes who immigrated here from Marxist torn Europe began their infection of American academia with socialist strategy, cultural Marxism, and political correctness was ridiculous and it was destroying our country. Yes he deserves credit for rescuing the narrative, for pointing out what we had all been fighting during the Obama years but the people – the movement – was undeniably ripe.

The other group, while grateful for Donald Trump's courage and tenacity to step in and knock out 17 primary opponents while never wavering in his ability to eschew the America First principles of protected borders, free trade, and an eradication of radical Islamic terrorism, recognizes the man as the messenger, not the movement embodied. There is a powerful common denominator but a very sticky numerator, to use a metaphor.

What makes the difference in mindset so fascinating is that the first group tends to adore the man regardless of his choices and decisions in office. This is antithetical to the very nature of a free society because it precludes any Trump election supporter from expressing any form of disagreement with the President in my opinion. The second group sees anyone in office as responsible for actions and obligated not only to keep campaign promises but also to explain when deviations occur. This is one of the reasons the Founders of our US Constitution wanted to safeguard our freedom to speak. Inherent in the First Amendment is the ability to "petition the government for a redress of grievances." Every office holder is accountable, including President Trump who we will forever be most grateful for busting up the establishment.

A Sea of New Media

The New Media by way of this sea of new media will advance the movement to move the country back toward the right, toward individual rights and smaller government and it will require all hands on deck. New Media related superpacs and coalitions dedicated to bringing down the establishment elites on both sides of the political aisle are forming everyday. This movement will bypass any one person, including President Trump. In the 2017 Virginia Primary gubernatorial race when President Trump endorsed the establishment candidate Ed Gillespie who was expected to trounce challenger Corey Stewart. In a stunning show of defiance at Trump's swamp endorsement, Stewart came within points of beating Gillespie. The swamp sat up and took notice. A few months later President Trump endorsed another establishment candidate named Luther Strange instead of pro-Trump candidate, Roy Moore in the Alabama Senate Primary. It was the sea of new media by virtue of social media that galvanized an already fired up America First base to over-ride the President's endorsement. In spite of establishment crony Majority Leader Mitch McConnell's $32 million dollar push to get establishment candidate Luther Strange to the finish line, the people of Alabama, convicted in their desire to drain the establishment swamp, voted overwhelmingly for Moore. One of the benefits of an awakened society that is now running much of its own media communications is independent thinking and decision-making. Needless to say, this was a game changing result that demonstrated that it's about the movement, not a single man.

How You Can Be The New Media

Your personal cell phone is a global megaphone with the power to defeat deceptions perpetrated by the mainstream media. Your cell phone is your primary tool. The most important thing you can do is to get completely comfortable with all the video graphic and audio graphic capability. What this means is that you must explore and practice with every video and photographic aspect of your camera. Your camera includes still shot photos, panoramic photos which can capture a span of 180° views, as well as live video. Within each of these capabilities you will find a lot of detailed features. The next thing you'll need to become very versed at is the use of key social media applications. These are additional programs that can be downloaded from the application store onto your cell phone. The two primary social media platforms are Twitter and Facebook. These are the places where you will set up accounts with your profile. These platforms are your channel to the outside world. They are the media through which you will disseminate your pictures, your thoughts, and your videos. Another primary platform is called Periscope. The Periscope application allows you to live stream, or video in real time, any event you are involved in and it allows you to do that in a forward facing or flipped to self-facing view in on the screen. Furthermore applications such as Twitter and Facebook are fully capable of being linked to streaming applications such as Periscope.

It is important to remember that all of these applications and platforms require some time investment in learning how to use them and getting proficient especially if you want to be able to capture moments in real time. It is relatively easy to learn and there are numerous resources online to help you navigate the instructions. Once you have become comfortable using them either as stand-alones or integrated with each other, all of it will become second nature. The ability to use your cell phone camera or video in the moment will allow you to interview people, capture notable events live, provide truthful evidence of an occurrence, and refute deceptions and mis-reports by others.

Where is New Media headed?

Power is shifting. New media does not need mainstream media for credibility, visibility, or to achieve distribution. If the mainstream media is really watching, smart they would start devising their plan to move into the social media sphere more aggressively. They would morph into their version of a social media presence. Mainstream media represents the behemoth of cable news agenda driven conglomerates and these entities are like huge ocean liners in that are difficult to turn around or course correct quickly. Mired in their own self interest & heavy bureaucracies they will likely never catch up to the tectonic shifts happening right now in social media's explosion. In fact by the time they figure it out and come to a consensus on it, they will be choking on the dust in the road.

In the meantime, we can expect New Media through social media to expand on every level. The primary focus of this expansion will be through the emergence of new platforms and the consolidation of existing ones such as Facebook, Instagram, Snapchat, Twitter, Pinterest, Youtube, and LinkedIn. These platforms will continue to develop as the conduits for expression both real and for augmented reality and they will consolidate their customer base and functions. The competition among platforms will increase noticeably around add-on capabilities in an attempt to become the go to platform for all forms of expression and commerce.

People are responding to the opportunity to create live content and this will lead platforms to develop more broadcast partnerships and connections. Another growing

A Sea of New Media

aspect of creating live content that should not be underestimated is the interactive nature of visual and audio projects. This by definition is a potential substitute for mainstream media's traditional on the ground interviewing and reporting. As more and more citizen journalists hit the ground to capture an event, they can conduct investigations and get the information out to the public, often faster than traditional outlets. While this version of the news may not always be accurate or may require additional outside vetting, it is no less reliable than the deception perpetrated on the public by the mainstream media. This trend will be very important to watch as technology explodes to support the needs and convenience of the citizen journalists.

On the consumer side it is clear to that there will be an ever-increasing number consumer users. The primary consumer platform tool, as Steve Jobs once predicted will continue to be the cell phone and application design will explode in response to this phenomenon. It makes sense that the smallest, most convenient way to carry so many functions and tools for expression and communication would be the personal cell phone.

The ability to monetize individually created content has opened up an entirely new avenue for doing business. Citizen journalists, bloggers, new media network developers are cashing in on what mainstream media has which is that content is valuable. Interesting content coupled with exclusivity and a good social media following commands a monetary reward. Citizen journalists have long sold their content to major outlets but now the trend is rapidly shifting to delete the middle-man. Individual content creators are generating income directly through the many platforms. This will be a huge trend to watch.

Recall "Pragmatic Primer for Realistic Revolutionaries?" Yes, I humbly borrowed the title from Andrew Breitbart's book, "Righteous Indignation: Excuse Me While I Save the World." Thank you Andrew, for conceiving the path ahead for the destruction of the Institutional Left and the resurgence of America. I humbly offer this summary of his priceless direction and leadership notes. In the chapter by the same name he lists the comparable list of aspirational actions by those wanting to join the conservative, economic nationalism, Goldwater conservative conscience values as the left has in Alinsky's Rules for Radicals. In effect, Andrew's list is Rules for Radicals on the right (non-violent). Here they are in his order:

What's coming is so much broader and deeper than just a morphing of traditional media. We can expect to see more think tanks for non-establishment policy, media, and cultural philosophy. New institutions will be emerging and some will be replacing old ones. This Sea of New

Media will continue to usher in a new era in accountability: in media, in politics, and in governments. The insemination has already begun for new political labels and new political parties both in substance and structure. Previously the news and life narratives could be controlled by a cartel of corporate heads, however the playing field has completely changed rendering the public much less vulnerable to the tactics of identity politics, political correctness, and fabricated narratives.

Grab your cell phone…we are all the media now.

Become a Force Multiplier

- Don't be afraid to go into enemy territory…
- Expose the Left for who they are—in their own words
- Be open about your secrets
- Don't let the Complex use its PC lexicon to characterize you and shape the narrative
- Control your own story—Don't let the Complex do it
- Engage in the social arena
- Don't pretend to know more than you do
- Don't let them pretend to know more than they do
- Ridicule is man's most potent weapon
- Truth isn't mean. It's truth.
- Believe in the audacity of hope.

https://www.reddit.com/r/The_Donald/comments/6umypa/
andrew_breitbarts_pragmatic_primer_for_realistic/
Accessed online September 13, 2017:

Made in the USA
Columbia, SC
28 November 2017